Copyright © 2023 Sam Steed

Contents

Quantumania Is Real

•••

By Sam Steed

•••

Here is What You Need to Know

Chapter 1: The Quantum Realm Frontier Unveiled

The Quantum Realm is an enigmatic and complex realm that has captivated the imagination of scientists and researchers for over a century. Since its discovery, the Quantum Realm has challenged our understanding of the laws that govern the universe and has opened a gateway to a world of infinite possibilities. In this article, we will embark on a journey to explore the world of quantum mechanics and uncover the mysteries of the Quantum Realm Frontier. We will delve into the history of the field, investigate the fundamentals of quantum mechanics, and examine the latest technological advancements that are shaping the future of science and technology. Join us as we unveil the mysteries of the Quantum Realm.

Introduction to the Quantum Realm

The quantum realm, also known as the quantum world, is a term used to describe the subatomic world of particles that make up the universe. This realm operates on a set of laws that are distinctly different from those of classical physics. At the quantum level, particles behave in ways that can be described as both particles and waves, and their behavior is often unpredictable.

Defining the Quantum Realm

The quantum realm is the smallest observable level of the physical universe. It is the world of particles that make up atoms and molecules, and it operates on a set of laws that are different from those of the classical physics that we are used to. At this level, quantum mechanics, or the rules that govern the behavior of subatomic particles, takes over.

Importance of Understanding the Quantum World

Understanding the quantum world is critical to understanding the universe we live in. It has significant applications in fields such as technology, medicine, and energy. Many modern-day technologies, including smartphones and computers, rely on the principles of quantum mechanics. Additionally, the study of quantum mechanics has led to significant advancements in areas such as cryptography and quantum computing, which have the potential to revolutionize the way we interact with technology.

Historical Development of the Quantum Theory

Early Ideas and Theories

The study of the quantum world can be traced back to the early 20th century when scientists began to realize that the classical physics that governs the macro world was insufficient in explaining the behavior of subatomic particles. Scientists like Max Planck and Albert Einstein started exploring the nature of the atom, and their insights led to the development of quantum theory.

Quantum Mechanics Pioneers

The pioneers of quantum mechanics, including Niels Bohr, Werner Heisenberg, and Erwin Schrödinger, made significant contributions to the development of quantum theory. They developed a set of principles that could describe the behavior of subatomic particles and demonstrated that the principles of classical physics were insufficient at this level.

The Fundamentals of Quantum Mechanics

Wave-Particle Duality

One of the fundamental principles of quantum mechanics is wave-particle duality. This principle states that particles can exist as both waves and particles, and their behavior depends on how they are observed.

Quantum Entanglement

Quantum entanglement is the idea that particles can become connected in such a way that the state of one particle affects the state of the other, no matter how far apart they are. This principle has significant implications for telecommunications and quantum computing.

Superposition

Superposition is the idea that particles can exist in multiple states at once. This principle is the foundation for the development of quantum computing.

Uncertainty Principle

The uncertainty principle is the idea that it is impossible to know the exact position and momentum of a particle at the same time. This principle has significant implications for the measurement of particles and the accuracy of scientific experiments.

The Quantum Realm Frontier Explored

Quantum Computing

Quantum computing takes advantage of the principles of quantum mechanics, including superposition and entanglement, to process information. Quantum computers have the potential to revolutionize the computing industry by solving complex problems and performing calculations much faster than traditional computers.

Quantum Teleportation

Quantum teleportation is the process of transmitting the quantum state of one particle to another particle, no matter how far apart they are. This process has significant implications for telecommunications and the transmission of information.

Quantum Cryptography

Quantum cryptography is a type of cryptography that relies on the principles of quantum mechanics to secure communication. It uses the principles of superposition and entanglement to transmit information securely and has the potential to revolutionize the field of cybersecurity.

The Future of Quantum Mechanics

Quantum mechanics, the branch of physics that studies the behavior of matter and energy at the atomic and subatomic levels, has made significant advancements in recent years. The future of quantum mechanics looks promising, with new technologies being developed and explored.

Quantum Technology Advancements

One of the most significant advancements in quantum technology is the development of quantum computers. Unlike classical computers that use bits to represent information, quantum computers use qubits, which can exist in multiple states simultaneously. This allows quantum computers to perform complex calculations faster and more efficiently than classical computers.

Another exciting advancement in quantum technology is the development of quantum sensors. These sensors can detect tiny changes in the environment and are incredibly accurate. They have already been used in various applications, such as detecting underground water sources and monitoring brain activity.

Potential Future Applications

The potential applications of quantum mechanics are vast, from improving cybersecurity to developing new materials and medicines. Quantum computers could revolutionize drug discovery by simulating complex chemical reactions and predicting their outcomes accurately. In addition, quantum sensors could greatly improve medical imaging, allowing doctors to see the body's internal structures with unparalleled resolution.

Applications of Quantum Mechanics in Modern Science

Quantum mechanics has already made significant contributions to modern science, with groundbreaking discoveries in various fields.

Quantum Physics in Medicine

One of the most well-known applications of quantum mechanics in medicine is magnetic resonance imaging (MRI). MRI machines use strong magnetic fields and radio waves to create detailed images of the body's internal structures. This technology is based on the principles of quantum physics and has revolutionized medical imaging.

Quantum mechanics is also being used to develop new cancer treatments. One promising approach is quantum dots, which are tiny particles that can be targeted to cancer cells, allowing for more precise treatment and fewer side effects.

Quantum Biology

Quantum mechanics is not just limited to the realm of physics and engineering. It also has implications in the field of biology. For example, recent research has suggested that quantum mechanics may play a role in the process of photosynthesis. Scientists believe that quantum effects enable plants to absorb energy from sunlight more efficiently.

Limitations and Challenges of Quantum Mechanics

Although quantum mechanics has made significant advancements, there are still limitations and challenges that need to be overcome.

Interpretation and Theory

The interpretation of quantum mechanics remains one of the most debated topics in physics. Despite the success of quantum mechanics in explaining the behavior of matter and energy at the atomic and subatomic levels, it is still unclear how it fits into a bigger picture of the universe.

Practical Limitations

One of the practical limitations of quantum mechanics is the difficulty in maintaining its fragile states. Quantum systems are susceptible to interference from the environment, which can cause them to lose their quantum properties. This is known as decoherence and is a significant challenge in developing practical quantum technologies.

Conclusion: The Quantum Frontier Unveiled

Quantum mechanics is a complex field that has the potential to revolutionize many aspects of our lives. From developing new technologies to advancing our understanding of the natural world, quantum mechanics has already made significant contributions to modern science. Although there are still limitations and challenges to overcome, the future of quantum mechanics looks very promising. In conclusion, the Quantum Realm has revealed to us an astonishing world of possibilities and has challenged our understanding of the universe. Through the efforts of pioneers in the field and advancements in technology, we have gained a deeper understanding of the Quantum Realm and its impact on modern science and technology. As we continue to explore the frontier of quantum mechanics, we can only imagine the possibilities that await us in the future.

Frequently Asked Questions (FAQ)

What is the Quantum Realm?

The Quantum Realm is a complex and enigmatic realm that deals with the behavior of matter and light on a subatomic level. It is governed by the laws of quantum mechanics, which differ significantly from classical physics.

What are the practical applications of quantum mechanics?

Quantum mechanics has a wide range of practical applications, from quantum computing and cryptography to medical imaging and drug discovery. It has the potential to revolutionize the way we live, work, and communicate.

What are the challenges associated with quantum mechanics?

One of the biggest challenges of quantum mechanics is the interpretation of the theory, which has been a subject of debate since its inception. Additionally, the practical limitations of current technology pose a challenge to the development of quantum technology.

Why is quantum mechanics important?

Quantum mechanics is important because it provides a fundamental understanding of the behavior of matter and energy on a microscopic level. It has been instrumental in the development of modern technology and has the potential to revolutionize fields such as medicine, energy, and communication.

Chapter 2: Quantum Weirdness and Everyday Life

Quantum mechanics is a highly complex and fascinating field of physics that has challenged even the greatest minds in science. It may seem like a purely theoretical concept, but the principles of quantum mechanics are present in our everyday lives. From the technology we use to the materials we interact with, quantum mechanics plays a vital role in shaping the world around us. In this article, we will explore some of the most intriguing aspects of quantum mechanics and its relationship with everyday life, from the double-slit experiment to quantum computing, and beyond.

Introduction to Quantum Weirdness

Quantum mechanics is a branch of physics that explains the strange behavior of particles at the quantum level that cannot be explained by classical physics. Quantum mechanics governs the behavior of the smallest particles in the universe, such as atoms, electrons, and photons.

Despite being more than a century old, quantum mechanics still baffles scientists and laypersons alike with its bizarre, counterintuitive, and seemingly impossible predictions.

What is Quantum Mechanics?

Quantum mechanics is a fundamental theory of physics that describes the behavior of particles at the quantum level. It was first introduced in the early 20th century to explain the properties of light and matter at the smallest scale.

In classical mechanics, particles are seen as solid, tangible objects with precise positions and momentum. However, quantum mechanics describes particles as wave-like entities that exist in multiple places simultaneously and can only be defined by probabilities.

The Uncertainty Principle

The uncertainty principle, one of the most famous principles in quantum mechanics, states that it is impossible to simultaneously know the precise position and momentum of a particle. This means that the more accurately we know the position of a particle, the less we know about its momentum, and vice versa.

This principle has profound implications, as it challenges our classical understanding of cause and effect and suggests that the universe is inherently unpredictable.

The Double Slit Experiment

The double-slit experiment is one of the most famous experiments in quantum mechanics. It was first conducted in the early 19th century to show the wave-like nature of light, and it has since become a hallmark experiment in quantum mechanics, demonstrating the bizarre behavior of particles at the quantum level.

Overview of the Experiment

In the double-slit experiment, a beam of particles, such as electrons or photons, is fired through a barrier with two slits. On the other side of the barrier, a screen is used to detect where the particles land.

The bizarre result is that the particles appear to pass through both slits simultaneously and create an interference pattern on the screen as if they were waves.

Interpretations of the Results

There are numerous interpretations of the results of the double-slit experiment, and each interpretation has its philosophical implications.

For instance, some physicists interpret the results as evidence of the wave-particle duality of matter, while others argue that the particles are in a state of superposition, existing in multiple states simultaneously.

Superposition and Entanglement

Superposition and entanglement are two of the most bizarre and fascinating phenomena in quantum mechanics.

Explanation of Superposition

Superposition is the idea that particles can exist in multiple states simultaneously. For instance, an electron can be in two or more energy levels at the same time.

This idea is counterintuitive to our classical understanding of the world, where objects can only exist in one state at a given time.

Entanglement and Its Implications

Entanglement is the idea that particles can become linked in such a way that their properties are dependent on each other, even when separated by large distances. This is known as quantum entanglement.

Entanglement has profound implications for the future of communication and computing, as it allows for faster and more secure transmission of information.

Quantum Computing

Quantum computing is a rapidly evolving field that uses the principles of quantum mechanics to process information.

What is Quantum Computing?

Quantum computing differs from classical computing because it uses the properties of quantum mechanics, such as superposition and entanglement, to perform computations. This allows for the processing of vast amounts of data at a much faster rate than classical computers.

Advantages of Quantum Computing

Quantum computing holds the promise of revolutionizing many fields, such as cryptography, drug design, and climate modeling, due to its ability to efficiently perform complex calculations that are beyond the capacity of classical computers.

Despite the challenges in developing and scaling quantum computers, the potential benefits of quantum computing make it an exciting and rapidly growing field.

Applications of Quantum Mechanics in Everyday Life

Quantum mechanics deals with the behavior of particles at a subatomic level, and it may seem like something that only scientists in their labs would care about. However, quantum mechanics has several applications in our everyday lives.

Quantum Cryptography

One application of quantum mechanics that has been gaining popularity is quantum cryptography. Quantum cryptography is a way to secure communication by using the properties of quantum mechanics. It is considered to be one of the most secure ways to transmit information as it is impossible to intercept or eavesdrop on the transmitted message without being detected.

Quantum Sensors

Another application of quantum mechanics that has been gaining attention is quantum sensors. Quantum sensors can detect the smallest changes in physical properties such as magnetic and electric fields, temperature, and pressure. They have several applications in fields such as healthcare, environmental monitoring, and military and defense.

Quantum Imaging

Quantum imaging is another application of quantum mechanics that has the potential to revolutionize the way we see things. It uses the properties of quantum mechanics to create images with superior resolution and clarity compared to traditional imaging techniques.

The Future of Quantum Technology

Quantum technology has the potential to revolutionize several fields, and scientists are continually exploring its possibilities.

Potential Developments in Quantum Computing

One of the most promising applications of quantum mechanics is quantum computing. Quantum computers have the potential to be significantly faster and more powerful than traditional computers, and they can solve problems that are currently considered impossible to solve. Quantum computers could revolutionize fields such as medicine, finance, and data analysis.

Impact of Quantum Technology on Society

The impact of quantum technology on society is still uncertain, but it has the potential to be significant. From better communication systems to advanced medical treatments, quantum technology could bring several benefits to society. However, it could also pose new challenges, such as ethical considerations and the impact on the job market.

Ethical Considerations in Quantum Computing

Quantum computing has several ethical implications that need to be considered.

Quantum Computing and National Security

One of the major ethical implications of quantum computing is its impact on national security. Quantum computers could break current encryption methods that are used to secure sensitive data, and this could pose a significant threat to national security.

Ethical Implications of Quantum Cryptography

While quantum cryptography is considered to be one of the most secure ways of transmitting messages, it also raises ethical questions. For example, if only the wealthy and powerful can afford quantum cryptography, it could create a digital divide. Furthermore, there is a risk that quantum cryptography could be used for illegal activities, such as money laundering or terrorism.

In conclusion, quantum mechanics has several applications in our everyday lives, and its potential for future development is massive. However, with the benefits come ethical considerations that need to be addressed. In conclusion, the strange and often perplexing world of quantum mechanics has immense potential to revolutionize the way we live, work, and interact with each other. While the full extent of this potential is yet to be realized, the applications of quantum mechanics in everyday life have already begun to transform the world around us. As we explore the possibilities of this exciting field, one thing is certain: the future is quantum, and it's full of endless possibilities.

Frequently Asked Questions

What is the double-slit experiment?

The double-slit experiment is a classic experiment in quantum mechanics that involves shining a beam of particles, such as electrons or photons, through two narrow slits and observing the resulting pattern on a screen. The experiment showcases the phenomenon of wave-particle duality and has significant implications for our understanding of quantum mechanics.

What is quantum entanglement?

Quantum entanglement refers to a phenomenon where two or more particles become linked in such a way that the state of one particle is dependent on the state of the other, even if they are separated by a large distance. This concept has been demonstrated through various experiments and has significant implications for quantum computing and cryptography.

What is quantum computing?

Quantum computing is a type of computing that uses quantum-mechanical phenomena, such as superposition and entanglement, to perform computations. This type of computing has the potential to solve certain problems much faster than classical computers and has applications in areas such as cryptography, finance, and materials science.

What ethical considerations are associated with quantum computing?

Quantum computing has significant implications for national security, as the powerful computational capabilities of quantum computers could be used to break encryption and security systems. Additionally, the development of quantum technologies could have socioeconomic implications, as it could exacerbate existing inequalities between countries and individuals. As such, it is important to consider the ethical implications of quantum computing as the technology continues to develop.

Chapter 3: Quantum Computing: Unleashing the Power

Quantum computing, once a theoretical concept, is now becoming a reality. This technology has the potential to revolutionize various industries by allowing scientists, researchers, and engineers to solve problems at a scale that was previously impossible. Unlike traditional computing, quantum computing uses quantum bits, or qubits, which can exist in multiple states simultaneously, to perform calculations. As such, quantum computers have the ability to solve complex problems much faster than traditional computers. In this article, we will explore the basics of quantum computing, its applications in various industries, its potential impact on society, challenges and limitations, advancements, and the promise it holds for the future.

Introduction to Quantum Computing

What is Quantum Computing?

Quantum computing is a revolutionary technology that harnesses the power of quantum mechanics, the physics that governs the behavior of matter and energy at the atomic and subatomic levels. Unlike classical computers, which rely on binary bits to store and process information, quantum computers use quantum bits or qubits, which can exist in multiple states at the same time, allowing them to perform calculations at an exponentially faster rate.

History of Quantum Computing

The concept of quantum computing was first proposed by physicist Richard Feynman in the early 1980s, but it wasn't until the late 1990s that the first experimental quantum computers were built. Since then, the field has advanced rapidly, with significant breakthroughs in quantum algorithms, error correction, and hardware design. Companies like IBM, Google, and Microsoft are investing heavily in quantum computing research, and experts predict that quantum computers could have a transformative impact on various industries.

Why is Quantum Computing important?

Quantum computing has the potential to solve complex problems that are currently intractable for classical computers, such as simulating the behavior of molecules, optimizing logistical systems, and breaking cryptographic codes. It could also lead to the development of new materials, drugs, and energy sources, and help address some of the biggest challenges facing humanity, from climate change to disease. As such, quantum computing is a strategic priority for many governments and companies worldwide.

The Basics of Quantum Mechanics

The Principles of Quantum Mechanics

Quantum mechanics is the branch of physics that describes how the universe behaves at the atomic and subatomic levels. It is based on several principles, including superposition, which states that a quantum system can exist in multiple states at the same time, and uncertainty, which states that the position and momentum of a particle cannot be precisely determined at the same time. Other principles of quantum mechanics include wave-particle duality and the observer effect, which describes how the act of observing a quantum system can change its behavior.

The State of Quantum Systems

In quantum mechanics, the state of a system is described by a wave function, which contains information about the probabilities of different outcomes of measurements. When a measurement is made, the wave function collapses into a single state, which determines the outcome of the measurement. The state of a quantum system can be represented by a qubit, which can exist in a superposition of the 0 and 1 states or any combination thereof.

Quantum Entanglement

Quantum entanglement is a phenomenon in which two or more qubits become correlated in such a way that the state of one qubit is linked to the state of the other qubit, even when they are separated by large distances. This allows for the possibility of secure communication and the development of quantum networks, which could be used for quantum key distribution and other applications.

Quantum Computing vs Traditional Computing

Differences and Similarities between Quantum and Traditional Computing

One of the main differences between quantum and traditional computing is the way they process information. While classical computers use binary bits, which can only exist in one of two states (0 or 1), quantum computers use qubits, which can exist in multiple states simultaneously. This allows quantum computers to solve certain problems exponentially faster than classical computers. However, both quantum and traditional computing share some similarities, such as the need for algorithms and processing power to perform calculations.

Limitations of Traditional Computing

Classical computers are limited by the physical constraints of their hardware, such as the number of transistors and the speed of their processing units. This makes it difficult for them to perform certain calculations or process large amounts of data. Additionally, classical computers are subject to errors due to noise and interference, which can affect the accuracy of their results.

Advantages of Quantum Computing over Traditional Computing

Quantum computing offers several advantages over classical computing, including the ability to solve problems that are currently intractable for classical computers, such as simulating chemical reactions and optimizing complex systems. Quantum computing could also lead to the development of new technologies, such as quantum cryptography and quantum communication, which could revolutionize the way we transmit and secure information. However, quantum computing is still in its early stages and faces many challenges, such as scaling up the number of qubits and improving the stability of quantum systems.

Applications of Quantum Computing in Various Industries

Quantum Computing in Healthcare

Quantum computing could have a significant impact on healthcare by enabling more accurate simulations of biological systems, which could lead to the development of new drugs and therapies. Quantum computing could also improve medical imaging and diagnosis, by analyzing large datasets of medical images and identifying patterns that are difficult for humans to detect.

Quantum Computing in Finance and Banking

In finance and banking, quantum computing could be used for portfolio optimization, risk management, and fraud detection. Quantum computing could also enable faster and more accurate simulations of financial markets, which could help investors make better-informed decisions.

Quantum Computing in Cybersecurity

Quantum computing could revolutionize cybersecurity by enabling the development of quantum-resistant encryption algorithms and the detection of quantum hacking attempts. Quantum cryptography, which uses the laws of quantum mechanics to transmit secure information, could also provide a more secure alternative to traditional cryptography.

Quantum Computing in Energy and Environment

Quantum computing could be used to optimize energy systems, such as renewable energy grids and energy storage systems. This could help reduce carbon emissions and improve the efficiency of energy production and distribution. Quantum computing could also be used for environmental monitoring and prediction, by analyzing large datasets of environmental data and identifying patterns that could help mitigate the effects of climate change.

Potential Impact of Quantum Computing on Society

Quantum computing has the potential to revolutionize several industries, including finance, healthcare, transportation, and energy. With its ability to process vast amounts of data quickly and accurately, quantum computing can lead to increased economic growth, job creation, and better social outcomes.

Quantum Computing and Economic Growth

Quantum computing can help businesses make better decisions by analyzing large data sets and complex algorithms. The ability to process and analyze data faster and more accurately can lead to better product development, operational efficiency, and improved customer experience. This, in turn, can boost economic growth and increase revenues for businesses.

Quantum Computing and Job Creation

As the use of quantum computing grows, there will be an increased demand for professionals with specialized skills in quantum physics, computer science, and mathematics. This could create new job opportunities in research, development, and implementation of quantum computing.

Quantum Computing and Social Implications

Quantum computing can help solve some of the world's biggest social problems by simulating complex natural phenomena and predicting outcomes for environmental issues, climate change, and disease control. It can also lead to advancements in personalized medicine, which could improve healthcare outcomes for individuals.

Challenges and Limitations of Quantum Computing

While quantum computing holds tremendous promise, it faces significant technical, cost, and sustainability challenges that need to be addressed before quantum computing can reach its full potential.

Technical Challenges

One of the biggest technical challenges of quantum computing is creating stable quantum bits (qubits) that can be used for computing. Qubits are highly sensitive to external disturbances, making them difficult to control and maintain. Additionally, scaling up quantum computers to commercial size is still a challenge.

Cost Challenges

Quantum computing is expensive due to the high cost of building and maintaining quantum computers. As a result, it is currently limited to governments and large corporations. The cost of quantum computing needs to come down significantly before it can be widely adopted.

Sustainability Challenges

Quantum computing is energy-intensive and requires significant amounts of power to operate. This can lead to high carbon emissions and environmental concerns. The development of more sustainable and eco-friendly quantum computing technology is needed to address this challenge.

Future of Quantum Computing and Its Advancements

Quantum computing is still in the early stages of development, but significant advancements are expected in the coming years.

Quantum Computing in the Next Decade

In the next decade, quantum computing is expected to move from research labs to commercial applications. The development of more stable and scalable qubits will lead to the creation of larger and more powerful quantum computers that can be used to solve complex problems.

Quantum Computing Beyond the Next Decade

Beyond the next decade, quantum computing is expected to reach its full potential and become an integral part of our daily lives. It will likely lead to the development of new technologies and industries that we cannot yet imagine.

Quantum Computing and the Intersection with Other Technologies

Quantum computing is expected to intersect with other technologies, such as artificial intelligence (AI) and blockchain, to create new applications and industries. For example, quantum computing can improve the accuracy and efficiency of AI algorithms, and blockchain can provide secure and transparent storage of quantum data.

Conclusion: The Promise of Quantum Computing

Quantum computing has the potential to revolutionize several industries, leading to increased economic growth, job creation, and better social outcomes. While quantum computing still faces technical, cost, and sustainability challenges, significant advancements are expected in the coming years. The future of quantum computing is exciting, and it has the potential to transform our world in ways we cannot yet imagine. In conclusion, quantum computing undoubtedly holds a lot of promise for the future. Despite the challenges and limitations that exist, advancements are being made every day to overcome them. The potential applications of quantum computing in various industries are vast, and the impact on society could be significant. As we continue to unlock the power of quantum computing, it will be exciting to see how this technology will shape the future of our world.

Frequently Asked Questions (FAQ)

What is quantum computing and how does it work?

Quantum computing is a type of computing that uses quantum bits, or qubits, which can exist in multiple states simultaneously, to perform calculations. Unlike traditional computing, where each bit can only be in one state at a time, qubits can exist in both states, allowing for parallel processing and the ability to solve complex problems much faster.

What are some potential applications of quantum computing?

Quantum computing has the potential to revolutionize various industries, including healthcare, finance and banking, cybersecurity, and energy and environment. It can be used to optimize drug discovery, improve financial modeling and portfolio management, enhance data encryption, and optimize energy systems.

What are some challenges and limitations of quantum computing?

One of the biggest challenges of quantum computing is the issue of stability and scalability. Quantum computers are extremely sensitive to their environment and require a highly controlled environment to function properly. They are also very expensive to build and maintain, and there are currently only a few quantum computers available worldwide. Additionally, not all problems can be solved using quantum computing, and it is still a relatively new technology that is being researched and developed.

What does the future of quantum computing look like?

The future of quantum computing is exciting and holds a lot of promise. There are many advancements being made in the field, and it is expected that quantum computers will become more powerful and accessible in the coming years. As such, quantum computing will likely have a significant impact on various industries and society as a whole.

Chapter 4: The Quantum Universe: Multiverse or Many Worlds?

The world of quantum mechanics has always been a fascinating subject for both scientists and philosophers. It challenges our conventional understanding of reality and raises questions about the true nature of our universe. One of the most intriguing topics in this field is the concept of multiverse and many world interpretations. In this article, we will explore these ideas in depth and compare them to each other. We will discuss the evidence that supports them and also examine their implications for science and philosophy. Join us on this journey as we dive into the mysterious quantum universe and try to make sense of its complex and mind-bending concepts.

Introduction to the Quantum Universe

Exploring the Mysteries of the Universe

The universe is an intriguing and complex phenomenon that has fascinated scientists and philosophers for centuries. With the advent of modern physics, we have been able to explore the universe in more depth than ever before. The study of quantum mechanics has revealed a plethora of new ideas and theories about the universe that are both fascinating and, at times, difficult to comprehend. In this article, we will delve into one of the most fascinating topics in the world of quantum mechanics: the multiverse.

The Concept of Multiverse

Understanding the Concept of Multiverse
Types of Multiverse

The multiverse is a fascinating concept that has been explored in various scientific fields, including physics and philosophy. The idea of the multiverse suggests that there might be multiple universes besides our own, each with its own unique set of physical laws and characteristics. This concept is based on the idea that our universe is not the only one that exists, but rather one of many possible universes that could exist.

There are various types of multiverse theories, including the bubble universe theory, the landscape multiverse theory, and the many worlds' interpretation of quantum mechanics. Each of these theories offers a unique perspective on the concept of the multiverse and provides insight into the possible existence of multiple universes.

Understanding Quantum Mechanics

Basic Concepts of Quantum Mechanics

Quantum Entanglement and Superposition

Quantum mechanics is a branch of physics that explores the behavior of matter and energy on the atomic and subatomic levels. The basic concepts of quantum mechanics include wave-particle duality, uncertainty principle, and quantum tunneling.

Quantum entanglement is a phenomenon that occurs when two particles become linked in such a way that the state of one particle is dependent on the state of the other. Superposition is a state in which a particle can exist in multiple states at the same time.

These concepts are crucial to understanding the many world's interpretations of quantum mechanics and the possibility of the multiverse.

Many Worlds Interpretation of Quantum Mechanics

What is Many Worlds Interpretation?
Criticism and Controversy

The many worlds interpretation of quantum mechanics suggests that every quantum event creates a new universe, resulting in multiple universes with different outcomes. In essence, every possible outcome of an event happens in a parallel universe.

While this theory is intriguing, it has also been met with criticism and controversy. Some scientists argue that it is impossible to verify the existence of other universes, making the theory untestable. Others question the validity of considering every possible outcome as a separate universe.

Despite the controversy surrounding the many worlds' interpretation, it remains a popular and thought-provoking theory in the world of quantum mechanics.

Comparison between Multiverse and Many Worlds

The concept of the multiverse and many worlds are two fascinating explanations for the quantum universe. The multiverse proposes the existence of a vast number of universes, each with different properties and laws of physics. On the other hand, the many-worlds interpretation suggests that every possible outcome occurs in a separate branch of the universe.

Similarities and Differences

Both theories propose the existence of multiple realities or universes. In both cases, these universes are inaccessible to us due to technological limitations. However, the difference lies in the explanation of these universes' existence. The multiverse theory is based on the idea of cosmic inflation, suggesting the creation of multiple universes during the big bang. In contrast, the many-worlds interpretation is based on the idea that quantum events split the universes into different branches.

Which One is More Plausible?

Neither one can be entirely proven or disproven yet, and both have their strengths and weaknesses. The multiverse theory provides a natural explanation for the complexity and diversity of our universe, while the many-worlds interpretation suggests a consistent explanation for the quantum theory's puzzling aspects. However, it is challenging to determine which is more plausible since both approaches are theoretical and have yet to be fully tested.

Evidence Supporting Multiverse and Many Worlds

Scientific Evidence

Empirical data cannot directly prove the existence of the multiverse or many worlds. However, some experiments and observations indirectly support these theories. For instance, the cosmic microwave background radiation and the distribution of galaxies support the cosmic inflation theory, while quantum experiments support the many-worlds interpretation.

Philosophical Arguments

Philosophical arguments in favor of the multiverse theory point out that it offers a plausible solution to the problem of fine-tuning. In contrast, arguments in favor of the many-worlds interpretation highlight that it provides a sensible explanation for the measurement problem in quantum mechanics.

Implications of Multiverse and Many Worlds for Science and Philosophy

Impact on Scientific Theories and Concepts

The multiverse and many worlds have implications for various scientific theories and concepts. They affect our understanding of cosmology, astrophysics, quantum mechanics, and even the nature of reality itself.

Philosophical Implications and Debates

The multiverse and many-worlds interpretation also has substantial philosophical implications. They affect our understanding of the meaning of existence, the concept of causality, and the limits of scientific knowledge.

Conclusion: The Future of Quantum Universe Research

Promising Areas of Research

The multiverse and many-worlds interpretation are still unproven and remain theoretical concepts. However, quantum universe research is continually evolving, and new observations and experiments may provide more evidence for one theory over the other. Some promising areas of research include testing quantum entanglement and analyzing cosmic microwave background radiation.

Final Thoughts on Multiverse and Many Worlds

The multiverse and many-worlds interpretations are intriguing and mind-bending theories. They provide fascinating explanations for the complexity and diversity of our universe and present interesting philosophical and scientific implications. It remains to be seen which, if either, will be proven right, and the future of quantum universe research is exciting. In conclusion, the concepts of multiverse and many worlds interpretation provide fascinating and thought-provoking insights into the nature of our reality. While much remains unknown and debated, the evidence and philosophical implications they present continue to intrigue and challenge researchers in the field of quantum mechanics. As we continue to explore the universe at both the macroscopic and microscopic levels, the theories and ideas presented in this article will undoubtedly play an important role in advancing our understanding of the quantum universe.

FAQ

What is the difference between the multiverse and many worlds' interpretations?

The multiverse theory suggests that there are multiple universes existing simultaneously, each with its unique properties. On the other hand, many worlds interpretation proposes that every time a quantum measurement is made, the universe splits into different versions, creating multiple worlds, each with a different outcome.

Is there any evidence supporting the multiverse and many worlds' interpretation?

While the evidence is not conclusive, some observations support the idea of a multiverse. For example, cosmic microwave background radiation has been found to have unusual patterns that could be explained by the presence of other universes. Similarly, the many worlds interpretation is supported by certain quantum experiments, such as the double-slit experiment.

What are the philosophical implications of the multiverse and many world interpretations?

Multiverse and many-worlds interpretations raise important questions about the nature of reality and our place in the universe. They challenge our conventional understanding of causality and determinism and suggest that the future is not predetermined. Moreover, they raise questions about the meaning of existence and the possibility of infinite versions of ourselves and our choices.

Are there any criticisms of the multiverse and many worlds interpretation?

Yes, there are many criticisms of both theories. Some physicists argue that there is no way to test the existence of other universes, making the multiverse theory unscientific. Similarly, the many worlds' interpretation has been criticized for its lack of clarity and the difficulty of reconciling it with our macroscopic experience. Critics also argue that it violates Occam's razor, which states that the simplest explanation is often the best.

Chapter 5: Quantum Entanglement: The Spooky Connection

Quantum entanglement, one of the most perplexing phenomena in modern physics, is often described as a "spooky connection" between two particles that allows them to behave as if they are correlated even when they are separated by vast distances. Despite decades of research, scientists still do not fully understand this mysterious phenomenon. In this article, we will explore the fascinating world of quantum entanglement, from its historical roots to its modern-day applications in technology. We will also delve into the challenges that arise in maintaining entangled states and the exciting future of quantum entanglement research.

Introduction to Quantum Entanglement

Quantum entanglement is one of the most fascinating phenomena in the world of physics. It involves a strange and mysterious connection between two particles that are separated by any distance, such that the state of one particle instantly affects the state of the other. This connection is so strong that it appears to violate the laws of classical physics.

What is Quantum Entanglement?

Quantum entanglement is a phenomenon where two particles are connected in such a way that the state of one particle is immediately affected by the state of the other, regardless of their distance from each other. This connection is a result of their creation in a way that their properties, like spin, are interlinked. Even when the particles are separated by a large distance, if one of them is measured, the other particle's state can be determined.

The Historical Context of Quantum Entanglement

Quantum entanglement has its roots in the early 20th century when physicists were trying to understand the behavior of subatomic particles. It was first proposed by Albert Einstein, Boris Podolsky, and Nathan Rosen, in a paper published in 1935. They used the term "entanglement" to describe the phenomenon of two particles interacting in such a way that their properties were linked.

Einstein's "Spooky Action at a Distance"

Einstein's Objection to Quantum Entanglement

Einstein was skeptical of quantum entanglement and referred to it as "spooky action at a distance." He believed that the phenomenon violated the principle of locality, which states that an object's properties cannot be influenced by anything outside its immediate surroundings. For Einstein, this was a serious problem, as it challenged the foundations of modern physics.

The EPR Paradox

In 1935, Einstein, Podolsky, and Rosen proposed the EPR paradox, which was based on the idea that entanglement could be used to transmit information faster than the speed of light, which was impossible according to the theory of relativity. The paradox was resolved by John Bell in the 1960s, who showed that the correlation between entangled particles couldn't be explained by classical physics.

Understanding Entangled States

Mathematical Description of Entangled States

Entangled states are described mathematically using a concept called superposition. When two particles are entangled, the state of the system can be described as a combination of both particles' states. The two particles are described by a single quantum state, which is known as an entangled state.

Types of Entangled States

There are different types of entangled states, such as Bell states, Werner states, and GHZ states. Bell states are the most famous ones and are used in quantum cryptography. Werner states are used to study the effects of entanglement in noisy environments, while GHZ states are used to test the limits of quantum mechanics.

Bell's Theorem and Violation of Local Realism

Local Hidden Variables Theory

Local hidden variables theory is a theory that assumes that there are hidden variables that determine the properties of particles, which are independent of any measurement. This theory was proposed to explain the correlations between entangled particles without violating the laws of classical physics.

Bell's Theorem and Experimental Verification of Quantum Entanglement

Bell's theorem states that any theory that satisfies certain reasonable assumptions must obey certain inequalities, which have been subsequently violated by experiments. This means that the local hidden variables theory is wrong, and quantum mechanics is the only way to explain the phenomenon of quantum entanglement. Experimental results have consistently confirmed Bell's theorem and the reality of quantum entanglement.

Applications of Quantum Entanglement in Technology

Quantum entanglement may sound like something out of science fiction, but it has a variety of applications in modern technology. Two of the most exciting areas of research are quantum cryptography and quantum computing.

Quantum Cryptography

Quantum cryptography is a method for secure communication over a network. Traditional cryptographic systems rely on complex mathematical algorithms, but these can be broken with enough computing power. Quantum cryptography, on the other hand, uses the principles of quantum entanglement to create unbreakable codes.

When two particles are entangled, their properties are intertwined in a way that is impossible to replicate. If someone tries to intercept a message encrypted using quantum cryptography, they would destroy the entangled state of the particles and the message would be irrecoverable. This makes quantum cryptography an extremely secure method for transmitting sensitive information.

Quantum Computing

Quantum computing is another area where quantum entanglement is being used to revolutionize technology. A quantum computer uses qubits, which can be in a superposition of states, rather than classical bits which can only be 0 or 1. This allows quantum computers to perform certain calculations much faster than classical computers.

One of the key challenges in building a quantum computer is maintaining the entangled state of the qubits. Any interaction with the environment can cause the entangled state to break down, a phenomenon known as decoherence. Researchers are exploring ways to mitigate these effects to build more powerful quantum computers.

Challenges in Maintaining Quantum Entanglement

While quantum entanglement has the potential to unlock new technologies, it also presents significant challenges. Two of the biggest challenges are decoherence and noise.

Decoherence

Decoherence occurs when the entangled state of two particles is disrupted by interactions with their environment. This can be caused by a variety of factors, including temperature changes, electromagnetic radiation, and physical disturbances. Once the entangled state is lost, it cannot be recovered, rendering any calculations or communications dependent on the entanglement useless.

Researchers are exploring ways to reduce the impact of decoherence, including isolating the entangled system from its environment and using error-correcting codes to preserve the information.

Noise and Disturbances

In addition to decoherence, noise, and disturbances can also disrupt the entangled state of particles. For example, if one particle becomes entangled with another due to an accidental interaction, this can interfere with the intended entanglement and cause errors in communication or computation.

Researchers are working to develop techniques to filter out unwanted entanglements and disturbances, such as using entanglement swapping to create more robust entangled states.

Future Directions in Quantum Entanglement Research

As researchers continue to unravel the mysteries of quantum entanglement, new applications and technologies are likely to be developed. Two areas of particular interest are quantum teleportation and quantum communication networks.

Quantum Teleportation

While it may sound like something out of Star Trek, quantum teleportation is a real phenomenon that has been demonstrated in experiments. Quantum teleportation involves transferring the quantum state of one particle to another, without physically moving the particle itself. This has the potential to revolutionize fields like encryption and computing.

Researchers are exploring ways to scale up quantum teleportation to larger systems, including developing techniques for teleporting multiple qubits at once.

Quantum Communication Networks

Quantum communication networks would allow for secure transmission of information over long distances. This could have important applications in fields like finance and national security. However, building a quantum communication network requires solving several technical challenges, including developing reliable quantum repeaters and establishing trusted nodes in the network.

Research in these areas is ongoing, with the goal of one day creating a global quantum communication network that can securely transmit information across the world. Quantum entanglement has captivated the imagination of scientists and the public alike, as it challenges our understanding of the fundamental nature of the universe. As research in this field continues to progress, it is clear that quantum entanglement has immense potential for revolutionizing fields such as cryptography and computing. However, as we continue to explore this strange and mysterious phenomenon, we must also be mindful of the limitations and challenges that arise along the way. Despite the obstacles, the future of quantum entanglement research holds endless possibilities for unlocking the secrets of the universe.

FAQ

What is quantum entanglement?

Quantum entanglement refers to a phenomenon in which two particles become correlated in such a way that their properties are dependent on each other, even when the particles are separated by a great distance.

Why is quantum entanglement so important?

Quantum entanglement is important because it challenges our understanding of the fundamental nature of the universe. It has implications for fields such as cryptography and computing and has the potential for revolutionizing these industries.

What are some challenges in maintaining entangled states?

One of the biggest challenges in maintaining entangled states is decoherence, which is the loss of coherence in a quantum system. Other challenges include noise and disturbances that can disrupt the entangled state.

What are some applications of quantum entanglement in technology?

Quantum entanglement has a wide range of applications in technology, including quantum cryptography, quantum computing, and quantum communication networks. These technologies have the potential to transform the way we think about information security and computing power.

Chapter 6: The Quantum Revolution: Shaping the Future

The quantum revolution is upon us, and it is shaping the future in ways that were once unimaginable. Quantum theory, once a theoretical concept in the realm of physics, is now playing an increasingly important role in modern technology, from computing and communication to materials science and artificial intelligence. The power of quantum mechanics lies in its ability to manipulate matter and energy at the smallest scales, leading to revolutionary breakthroughs in fields such as cryptography and medicine. In this article, we will explore the quantum revolution and its significance, discussing various aspects of quantum theory and its applications that are shaping the future.

Introduction to Quantum Theory and Its Applications

Quantum theory is a fundamental branch of physics that has revolutionized our understanding of the world around us. It describes the behavior of matter and energy at the atomic and subatomic level, where classical mechanics fails to yield accurate predictions. The principles of quantum theory have led to the development of some of the most innovative technologies of our time, including quantum computing, quantum communication, quantum sensors, and metrology. In this chapter, we will explore the history and applications of quantum theory, as well as the future of quantum technology.

Quantum Mechanics: A Brief Overview

Quantum mechanics is a set of mathematical and conceptual tools that allow us to describe the behavior of microscopic particles such as electrons, protons, and photons. It is based on the principles of wave-particle duality, superposition, and entanglement, which are radically different from the classical physics we are familiar with. Quantum mechanics predicts that particles can exist in multiple states at once, that they can be entangled, and that their properties cannot be precisely measured at the same time. These principles have been experimentally verified and form the basis of a vast range of modern technologies.

The Birth of Quantum Theory and Its Evolution

Quantum theory was born in the early 20th century, when physicists such as Max Planck, Albert Einstein, and Niels Bohr began to study the behavior of atoms and the radiation they emitted. Their work led to the discovery of quantization, the idea that energy comes in discrete packets or quanta, and to the development of the wave-particle duality principle. In the following years, several other important concepts were introduced, including Heisenberg's uncertainty principle, Schrödinger's wave equation, and the concept of entanglement. Today, quantum theory is one of the most successful and widely used theories in all of science.

Applications of Quantum Mechanics in Modern Technology

Quantum mechanics has numerous applications in modern technology, including in the development of semiconductors, lasers, and transistors. It also underlies the principles of nuclear magnetic resonance (NMR) imaging and magnetic resonance imaging (MRI) used in medical diagnostics. In recent years, the principles of quantum mechanics have given rise to a new class of technological applications, including quantum computing, quantum communication, quantum sensors, and metrology. These technologies are poised to revolutionize the way we perform computing, communication, and measurement, paving the way for new scientific discoveries and breakthroughs.

The Significance of Quantum Computing

Quantum computing is one of the most exciting applications of quantum mechanics and promises to revolutionize computing as we know it. Unlike classical computers that use binary digits (bits) to represent information, quantum computers use quantum bits (qubits) that can exist in multiple states at once. This allows quantum computers to perform certain computations exponentially faster than classical computers, making them ideal for solving complex problems that are beyond the scope of classical computing.

Quantum Computing Principles and Mechanisms

In a quantum computer, qubits are manipulated using quantum gates to perform operations such as entanglement and superposition. These operations allow the computer to solve complex problems such as factoring large numbers and simulating complex molecules. However, quantum computing is still in its infancy, and building a reliable quantum computer that can perform useful computations remains a significant challenge.

Quantum Computing vs. Classical Computing

Classical computing is based on the principles of Boolean logic and uses bits to represent information. Quantum computing, on the other hand, uses qubits that can exist in multiple states at once, allowing quantum computers to perform certain computations exponentially faster than classical computers. While classical computers are still more practical for many everyday tasks such as word processing and internet browsing, quantum computing is poised to have a significant impact on scientific research, cryptography, and financial modeling.

Current Applications and Potential Uses of Quantum Computing

Quantum computing is still in the experimental stage, but researchers are already exploring its potential applications in fields such as finance, materials science, cryptography, and pharmaceuticals. Quantum computers could potentially be used to simulate complex chemical reactions, optimize financial portfolios, and break current encryption algorithms. As quantum computers become more powerful and reliable, their potential applications will likely continue to expand.

Quantum Communication and Cryptography

Quantum communication and cryptography are two of the most promising applications of quantum mechanics for secure communication and information exchange. These technologies are based on the principles of quantum superposition and entanglement, which enable secure communication that is virtually impossible to intercept or hack.

Quantum Key Distribution: A Secure Way of Communication

Quantum key distribution (QKD) is a protocol for secure communication that uses the properties of quantum mechanics to protect information exchange. QKD relies on the principle of entanglement to distribute keys that are used to encrypt and decrypt messages. Since the key is distributed using quantum mechanics, any attempt by an eavesdropper to intercept the key will be instantly detected, making QKD an ideal protocol for secure communication.

Quantum Cryptography: A New Era of Information Security

Quantum cryptography is a new paradigm for information security that uses the principles of quantum mechanics to protect data exchange. In quantum cryptography, the security of the data exchange is based on the laws of physics rather than complex mathematical algorithms. Quantum cryptography promises to provide unbreakable encryption, which is immune to attacks such as brute-force hacking and quantum attacks.

Challenges and Limitations of Quantum Cryptography

While quantum cryptography holds great promise for secure communication, there are still several challenges and limitations that need to be addressed before it can be widely adopted. Some of these challenges include the short range of quantum communication systems, the difficulty of maintaining entangled states over long distances, and the high cost of quantum hardware. However, with continued research and development, these challenges are likely to be addressed, paving the way for a new era of secure communication.

Quantum Sensors and Metrology

Quantum technology is also being used to develop new sensors and measurement methods that promise to provide greater accuracy and sensitivity than traditional techniques. These technologies are based on the principles of quantum interference and entanglement, which allow for highly precise measurements of physical quantities.

Quantum Sensors: A New Way of Measuring Physical Quantities

Quantum sensors are devices that use the principles of quantum mechanics to measure physical quantities such as magnetic fields, electric fields, and temperature. These sensors can achieve unprecedented levels of accuracy and sensitivity, making them ideal for applications such as navigation, mineral exploration, and medical diagnostics.

Quantum Metrology: Achieving Greater Accuracy with Quantum Technology

Quantum metrology is a field of study that aims to achieve greater accuracy in measurement using quantum technology. This involves using the principles of quantum mechanics to perform highly precise measurements of physical quantities such as time, length, and frequency. Quantum metrology promises to provide significant improvements in accuracy over traditional metrology techniques, making it an essential tool for scientific research and industrial applications.

The Future of Quantum Sensing and Metrology

Quantum Simulation and Materials Science

As we continue to explore the possibilities of quantum technology, one area that is gaining significant interest is quantum simulation. This technology has the potential to revolutionize the field of materials science by allowing us to simulate and test complex materials in a virtual environment, saving time, and resources.

Advancements in Quantum Simulation Techniques

Quantum simulation techniques have come a long way from their initial stages. Researchers are now using quantum computers to simulate complex chemical reactions and predict the behavior of new materials. This technique has already been used to design new catalysts for chemical reactions and develop new drugs more efficiently.

Quantum Materials Science: Solving Complex Problems with Quantum Computing

Quantum materials science is a field that combines quantum mechanics with materials science to solve complex problems that are difficult to approach using traditional methods. Quantum computers can be used to model the different properties of materials and simulate how they would behave under different conditions. This can lead to the discovery of new materials with unique properties that could be used in various industries.

The Role of Quantum Mechanics in Materials Science

Quantum mechanics is the study of the behavior of matter and energy at a microscopic level. It is a fundamental concept in materials science, as it helps scientists understand the different properties of materials, such as their strength, conductivity, and magnetism. Quantum mechanics is also essential in the development of nanotechnology, which involves manipulating materials on an atomic or molecular scale.

Advancements in Quantum Algorithms

Quantum algorithms form the backbone of quantum computing. They are used to solve complex problems that are beyond the capabilities of classical computers. As quantum technology continues to evolve, researchers are developing new algorithms that can solve problems faster and more efficiently.

Quantum Algorithm Design: A New Approach to Problem Solving

Quantum algorithm design is a new approach to problem-solving that takes advantage of the unique properties of quantum computers. Researchers are developing algorithms that can solve problems in optimization, cryptography, and machine learning, among others. One example of a quantum algorithm is Shor's algorithm, which can factor large numbers more efficiently than classical computers.

Quantum Error Correction: Overcoming the Challenges of Quantum Computing

One of the main challenges of quantum computing is dealing with errors. Errors can occur due to environmental factors or imperfections in the hardware. Quantum error correction involves identifying and correcting errors in quantum systems, making it possible to build more reliable quantum computers.

Applications of Quantum Algorithms in Various Fields

Quantum algorithms have the potential to impact various industries, such as finance, healthcare, and transportation. For instance, they can be used to optimize supply chains, develop new drugs, and improve traffic flow.

Quantum Machine Learning and Artificial Intelligence

Machine learning and artificial intelligence are two areas that could be transformed by quantum technology. Quantum computers have the potential to solve complex machine learning problems faster and more efficiently than classical computers.

Quantum Machine Learning: A New Era of Data Analytics

Quantum machine learning involves using quantum computers to process large amounts of data and make predictions. This technology has the potential to revolutionize data analytics by allowing us to process larger data sets and make more accurate predictions.

Quantum Artificial Intelligence: Enhancing Machine Learning with Quantum Technology

Quantum artificial intelligence is the integration of quantum computing with machine learning. It involves using quantum computers to speed up the training and optimization of machine learning models. This technique could lead to the development of more accurate and powerful machine learning models.

The Future of Quantum Machine Learning and AI

The future of quantum machine learning and AI is promising. Researchers are working on developing new algorithms and techniques that can take advantage of the unique properties of quantum computers. With more powerful quantum computers on the horizon, the possibilities for quantum machine learning and AI are endless.

Future Implications and Challenges for the Quantum Revolution

The quantum revolution is still in its early stages, but it has the potential to impact various industries and transform the way we approach complex problems. However, there are still challenges that must be overcome before we can fully realize the potential of quantum technology.

Current Limitations and Challenges of Quantum Technology

One of the main challenges of quantum technology is building and maintaining stable quantum systems. These systems are very delicate and are easily affected by environmental factors such as temperature and humidity. Another challenge is the limited number of qubits in current quantum computers, which limits their computational power.

The Potential Impact of Quantum Technology on Various Industries

The potential impact of quantum technology on various industries, such as finance, healthcare, and transportation, is significant. Quantum technology could lead to the development of new materials, drugs, and algorithms that could solve some of the world's most pressing problems.

Future Developments and the Road Ahead for Quantum Technology

The road ahead for quantum technology is promising. Researchers are working on developing more powerful quantum computers, better algorithms, and more stable quantum systems. With continued advancements in quantum technology, we could see a world where quantum computers are as ubiquitous as smartphones.In conclusion, the quantum revolution is rapidly transforming our world, opening new avenues for technological innovation and scientific discovery. While there are still many challenges and limitations to overcome, the potential benefits of quantum technology are vast and promising. With continued research and development, the quantum revolution is sure to unlock new frontiers and shape the future in ways that we can only begin to imagine.

Frequently Asked Questions

What is quantum theory?

Quantum theory is a branch of physics that describes the behavior of matter and energy at the smallest scales. It is based on the principles of quantum mechanics, which govern the behavior of subatomic particles and their interactions with electromagnetic radiation.

What is quantum computing?

Quantum computing is a form of computing that uses quantum-mechanical phenomena, such as superposition and entanglement, to perform operations on data. Unlike classical computers, which use binary digits (bits) to store and process information, quantum computers use quantum bits (qubits) that can exist in multiple states simultaneously.

How does quantum cryptography work?

Quantum cryptography uses the principles of quantum mechanics to transmit secure information over a communication channel. The method relies on the fact that any attempt to measure the quantum state of a photon (the particle used to transmit information) will change its state, making it impossible for an eavesdropper to intercept the message without being detected.

What are the challenges of quantum technology?

One of the main challenges of quantum technology is the issue of decoherence, which refers to the loss of quantum coherence, or the ability of qubits to exist in multiple states simultaneously. Decoherence can occur due to interactions with the environment, leading to errors and inaccuracies in quantum computing and other applications. Another challenge is the difficulty of scaling up quantum systems to handle larger amounts of data and perform more complex operations.

reality and opens up new possibilities for exploring the nature of existence.

Quantum Consciousness and Spirituality

Quantum consciousness has also been linked to spirituality and mystical experiences. Some researchers believe that the interconnectedness of particles in entanglement could be a potential explanation for spiritual experiences of unity and oneness. This opens up new avenues for exploring the intersection between science and spirituality.

The Philosophical Implications of Quantum Consciousness

The philosophical implications of quantum consciousness are far-reaching. It challenges traditional notions of the self, identity, and free will. It also raises questions about the nature of reality and the relationship between the observer and the observed. These questions could have profound implications for our understanding of ourselves and the world around us.

1. Practical Applications of Quantum Consciousness Research

Quantum Computing and AI

One of the most promising applications of quantum consciousness research is in the field of quantum computing. By harnessing the power of quantum entanglement, scientists hope to create computers that can perform complex calculations at lightning-fast speeds. This could lead to breakthroughs in fields such as artificial intelligence, cryptography, and chemistry.

Quantum Medicine

Quantum consciousness research also has implications for the field of medicine. Researchers are exploring the potential of quantum entanglement in developing new diagnostic tools and treatments for a wide range of diseases. This could lead to more personalized and effective treatments for patients.

Quantum Psychology

Quantum consciousness research could also have implications for the field of psychology. By studying the interconnectedness of particles in entanglement, researchers hope to gain new insights into the nature of consciousness and the workings of the mind. This could lead to breakthroughs in understanding mental health, perception, and cognition.

1. The Future of Quantum Consciousness Research and Its Potential Impact on Society

The Promise of Quantum Consciousness Research

The promise of quantum consciousness research is immense. It offers new ways of understanding ourselves and the world around us. It could lead to breakthroughs in fields such as computing, medicine, and psychology. It could also help to bridge the gap between science and spirituality, leading to a more holistic and integrated view of reality.

The Ethical Implications of Quantum Consciousness Research

As with any new field of research, there are ethical implications to consider. Researchers need to be mindful of the potential consequences of their work and ensure that their research is conducted ethically. This includes considering issues such as data privacy, informed consent, and potential unintended consequences.

The Future of Quantum Consciousness and Its Impact on Society

The future of quantum consciousness is exciting and full of possibilities. As our understanding of this field continues to grow, we may see breakthroughs in fields such as computing, medicine, and psychology. We may also see a new understanding of the nature of reality and our place in the universe. The implications of this research are vast, and it has the potential to transform our society in ways we cannot yet imagine. In conclusion, the intersection of quantum mechanics and consciousness opens up a whole new world of possibilities for understanding the mysteries of the universe and ourselves. While there is still much to be discovered and explored, the potential implications of quantum consciousness are immense, ranging from advancements in technology to a deeper understanding of the philosophical and spiritual aspects of our existence. As we continue to delve deeper into this field of research, we can only imagine what secrets and discoveries await us and the impact they will have on our understanding of the world around us.

FAQ

What is quantum consciousness?

Quantum consciousness refers to the theory that consciousness is a product of quantum processes in the brain. The theory suggests that consciousness arises from the entanglement of quantum particles in the brain, giving rise to the subjective experience of awareness.

What are the practical applications of quantum consciousness research?

The practical applications of quantum consciousness research are still being explored, but there is a great deal of potential for advancements in fields such as quantum computing, AI, and medicine. For example, quantum computers could be used to simulate complex quantum systems, leading to breakthroughs in fields such as drug discovery and materials science.

What is the relationship between quantum entanglement and consciousness?

Quantum entanglement is the phenomenon where two quantum particles become intertwined and share a state, even if they are separated by great distances. Some theories suggest that this phenomenon could be related to the nature of consciousness, as the entanglement of particles in the brain may give rise to subjective experience.

What are the ethical implications of quantum consciousness research?

As with any field of research, there are ethical considerations that need to be taken into account when exploring quantum consciousness. For example, there may be concerns around the use of new technologies that arise from this research, particularly if they involve the manipulation of consciousness. It will be important for researchers and society as a whole to consider the potential implications of quantum consciousness and ensure that any advancements in this field are used for the greater good.

Chapter 8: Bridging the Quantum Gap

Quantum computing has emerged as a potentially game-changing technology with the power to address complex problems that are beyond the capabilities of classical computing. Despite the immense potential of quantum computing, the technology remains in its nascent stages, and there are major hurdles that must be overcome to achieve its full potential. In this article, we will explore the challenges in quantum computing and discuss how researchers and engineers are working to bridge the gap between classical and quantum computing. We will also examine the potential impact of quantum computing in real-world applications, such as cybersecurity, finance, and drug discovery. Finally, we will consider the future of quantum computing and its implications for the computing industry as a whole.

Introduction to Quantum Computing

Defining Quantum Computing

Quantum computing is a type of computing that uses quantum phenomena, such as superposition and entanglement, to perform operations on data. Unlike classical computing, which relies on bits, quantum computing uses quantum bits, or qubits, which can exist in both the '0' and '1' states simultaneously, allowing for exponentially faster computations.

History of Quantum Computing

The concept of quantum computing was first proposed by physicist Richard Feynman in 1982, but it wasn't until the 1990s that the first qubits were successfully created and manipulated. Since then, advances in technology and research have brought quantum computing closer to reality.

Why Quantum Computing Matters

Quantum computing has the potential to revolutionize industries such as finance, medicine, and energy by solving complex problems that classical computing cannot handle. For example, quantum computing could help develop new drugs, optimize financial portfolios, and improve sustainable energy sources.

Challenges in Quantum Computing

The Complexity of Quantum Computing

Quantum computing is inherently complex and difficult to understand, even for experts. The math and physics behind it can be challenging, and the technology is not yet mature enough to be widely accessible.

Physical Limitations on Quantum Computing

Quantum computing requires extremely precise and stable conditions, and even the smallest interference can cause errors in computations. Additionally, the technology is currently limited by the number of qubits that can be reliably manipulated.

The Need for Specialized Knowledge

Quantum computing requires a whole new set of skills and knowledge that most computer scientists and engineers do not possess. Researchers and developers need to be familiar with quantum mechanics, error correction, and quantum algorithms.

Bridging the Gap between Classical and Quantum Computing

Classical vs. Quantum Computing: What's the Difference?

Classical computing relies on binary code represented in bits, whereas quantum computing relies on qubits, which can exist in multiple states at once. Additionally, quantum computing algorithms are usually probabilistic, meaning they produce results that are only statistically accurate.

Hybrid Computing: Combining Classical and Quantum Computing

One way to address the challenges in quantum computing is by utilizing classical computing in combination with quantum computing. Hybrid computing allows classical computers to handle tasks that are not well-suited for quantum computers, while quantum computers tackle computationally intensive tasks.

Quantum Computing as a Service

Another way to bridge the gap is by offering quantum computing as a service, similar to the way cloud computing works. This would allow researchers and businesses to access quantum computing resources without needing their own quantum computers or specialized knowledge.

Quantum Algorithms and Their Potential Impact

What are Quantum Algorithms?

Quantum algorithms are a set of instructions for performing operations on qubits. They are designed to take advantage of the unique properties of qubits to perform computations faster than classical algorithms.

Shor's Algorithm and its Implications for Cryptography

Shor's algorithm is a quantum algorithm that can factorize large numbers exponentially faster than classical algorithms. This has significant implications for cryptography, as many current security protocols rely on the inability to factor in large numbers.

Quantum Machine Learning

Quantum machine learning is a promising area of research that combines quantum computing and machine learning. It has the potential to improve pattern recognition, natural language processing, and other applications of machine learning.

Quantum Computing in Real World Applications

Quantum computing is no longer just a concept in a textbook. It is now being developed by companies such as IBM, Google, and D-Wave, with real-world applications that could revolutionize industries as diverse as cybersecurity, finance, and materials science.

Quantum Encryption and Cybersecurity

Quantum encryption is one of the most promising applications of quantum computing. Unlike traditional encryption methods, which are based on complex mathematical algorithms, quantum encryption uses the principles of quantum mechanics to encrypt information. This makes it extremely difficult for hackers to intercept and decode messages. As cybersecurity threats become more sophisticated, quantum encryption could become the gold standard for secure data transfer.

Quantum Computing in Finance

Quantum computing has the potential to revolutionize the finance industry by enabling the analysis of vast amounts of data in real time. This could have applications in areas such as risk assessment, portfolio optimization, and fraud detection. Some financial institutions such as JP Morgan, Barclays, and Goldman Sachs are already exploring how quantum computing can benefit their businesses.

Quantum Computing in Drug Discovery and Materials Science

Quantum computing could also have a profound impact on drug discovery and materials science. By simulating the behavior of atoms and molecules, quantum computers could help researchers to develop new drugs and materials, and to study the properties of existing ones. This could lead to more efficient drug discovery, the development of new materials with novel properties, and a better understanding of the world around us.

Future of Quantum Computing

While quantum computing is still in its infancy, there are already some exciting developments on the horizon that could change the game even further.

Advancements in Quantum Computing Hardware

One of the biggest challenges facing quantum computing is the development of hardware that can handle the complex computations required. Advances in quantum hardware, such as the development of qubits that are more stable and less error-prone, could pave the way for more powerful quantum computers that can handle more complex problems.

Quantum Computing and AI

Quantum computing and artificial intelligence are two of the most promising areas of technological development. By combining the two, researchers could create machines that are far more intelligent and capable than anything we have today. This could have a significant impact on fields such as medicine, finance, and even politics.

Quantum Computing Adoption and Accessibility

Finally, the accessibility of quantum computing is becoming an increasingly important issue for businesses and researchers. As more companies enter the market and develop quantum computing technologies, the cost of entry is likely to decrease. This could make quantum computing accessible to smaller businesses and research teams, enabling them to take advantage of its many benefits.

Conclusion and Implications for the Computing Industry

Quantum computing has the potential to revolutionize the computing industry in ways that we can't yet imagine. However, there are also significant challenges that need to be overcome before it can become a mainstream technology.

Challenges and Opportunities for Organizations

One of the biggest challenges facing organizations is adapting to the new paradigm of quantum computing. This will require significant investment in new technologies and processes, as well as a willingness to be flexible and adaptable.

The Impact of Quantum Computing on Computing as a Whole

The impact of quantum computing on computing as a whole is likely to be profound. It could enable us to solve problems that are currently unsolvable, provide new insights into the behavior of the universe, and usher in a new era of technological progress. However, it will also create new challenges and ethical considerations that we have yet to confront. Overall, the implications of quantum computing for the computing industry are both exciting and daunting, and it will be fascinating to see how this technology develops in the years to come. As quantum computing continues to evolve and become more accessible, its impact on the computing industry and beyond will only grow. While there are still many challenges to overcome, the potential benefits of quantum computing are too significant to ignore. It is an exciting time for quantum computing, and we look forward to seeing how this technology will transform the world in the years to come.

FAQs

What is quantum computing and how does it differ from classical computing?

Quantum computing is a type of computing that uses quantum bits or qubits instead of classical bits, which allows for the creation of algorithms that can solve problems that are beyond the capabilities of classical computers. While classical bits can represent either a 0 or a 1, quantum bits can represent both a 0 and a 1 at the same time, allowing for faster and more efficient computation.

What are some of the challenges in quantum computing?

One of the major challenges in quantum computing is the problem of decoherence, which occurs when quantum systems interact with their environment and lose their quantum properties. Another challenge is the difficulty of building and maintaining a large number of qubits, as any errors or noise can quickly compound and render the computation useless. Additionally, the need for specialized knowledge and expertise in quantum computing can make it difficult to find qualified researchers and engineers.

What are some real-world applications of quantum computing?

Quantum computing has the potential to revolutionize a wide range of industries, from finance and cybersecurity to drug discovery and materials science. For example, quantum computing could enable the development of more secure encryption methods that are resistant to hacking, or accelerate the discovery of new drugs and materials by simulating complex chemical reactions.

When will quantum computing become mainstream?

While quantum computing is still in its early stages, researchers and engineers are making rapid progress in developing new hardware and software tools that could bring quantum computing closer to mainstream adoption. However, it is difficult to predict when exactly quantum computing will become mainstream, as there are still many technical and practical challenges that must be overcome.

Chapter 9: Embracing the Quantumania World

The world is rapidly changing, and technology is at the forefront of this change. One of the latest technological advancements that has taken the world by storm is quantum computing. With its promise of unparalleled computing power and potential to revolutionize various industries, many have dubbed this new paradigm Quantumania. In this article, we will explore the basics of quantum computing, the current state of quantum technologies, and their potential applications in various industries. We will also discuss the benefits and challenges of embracing the Quantumania world and provide resources for those interested in getting involved in this exciting field.

Introduction to the Quantumania World

Welcome to the quantumania world, where science fiction meets modern reality. The quest for speed, accuracy, and efficient problem-solving has led us to the world of quantum computing. From tech giants to governments, everyone is racing towards harnessing the power of quantum computing to transform the way we approach complex problems. In this article, we'll introduce you to the world of quantumania, and explore the possibilities and challenges it brings.

Defining Quantumania

Quantumania refers to the growing enthusiasm and investment in quantum technologies, including quantum computing, quantum sensors, and quantum communication. Quantum mechanics is a branch of physics that describes the behavior of energy and matter at a very small scale. The principles of quantum mechanics have been applied in the development of quantum technologies, which have the potential to revolutionize fields such as cryptography, healthcare, and logistics, to name a few.

The Rise of Quantum Computing

Quantum computing is a type of computing that uses quantum phenomena to perform calculations. Quantum computers can solve certain problems exponentially faster than classical computers. The race for the development of quantum computers is driven by the potential to revolutionize fields such as cryptography, logistics, and drug discovery. Major tech companies, such as Google, IBM, and Microsoft, have invested heavily in quantum computing research and development, and have made significant progress in recent years.

Understanding Quantum Computing

To understand quantum computing, we need to understand the basics of quantum mechanics and how it differs from classical mechanics.

The Basics of Quantum Mechanics

Quantum mechanics describes the behavior of particles at the atomic and subatomic levels. It is characterized by the concept of superposition, in which a particle exists in multiple states simultaneously, and the principle of entanglement, which describes the correlation between the states of two particles. These principles allow for the development of quantum technologies that can process information at a scale not possible with classical computers.

How Quantum Computers Work

Quantum computers use qubits, the quantum version of classical bits, which can exist in multiple states simultaneously. These superpositioned states allow quantum computers to perform calculations exponentially faster than classical computers. However, quantum computing is still in its early stages, with challenges such as maintaining qubit stability and scaling quantum computers to commercial levels.

The Current State of Quantum Technologies

Quantum technologies are still in their early stages of development, but progress has been made in recent years.

Overview of Quantum Technology Development

Quantum technologies, such as quantum sensors and quantum communication, have seen significant developments in recent years. Quantum sensors can detect changes in gravity, magnetic fields, and temperature, and have applications in fields such as geology, navigation, and healthcare. Quantum communication allows for more secure and efficient communication than classical communication methods.

Big Names in Quantum Computing

Tech giants such as IBM, Google, and Microsoft have invested heavily in quantum computing research and development. Other companies, such as Rigetti Computing and D-Wave Systems, are also making significant strides in the field.

Applications of Quantum Computing in Various Industries

Quantum computing has the potential to revolutionize industries across the board.

Finance and Cryptography

Quantum computing could significantly impact the finance and cryptography industry. Quantum computing could break current cryptographic algorithms, leading to the need for new encryption methods. Quantum computing could also optimize financial calculations, such as portfolio optimization, fraud detection, and risk management.

Healthcare

Quantum computing could revolutionize drug discovery, enabling researchers to simulate and analyze complex molecular interactions. Quantum computing could also improve patient care by enabling faster diagnosis by analyzing large medical datasets.

Energy

Quantum computing could significantly impact the energy sector by optimizing energy production, transportation, and storage. Quantum computing could also enable more efficient and accurate weather forecasting, improving disaster preparedness.

Logistics and Transportation

Quantum computing could improve logistics and transportation by optimizing routing and scheduling and reducing travel times and emissions. Quantum computing can also improve inventory management and supply chain optimization.

The Future of Quantum Computing

In the realm of computing, quantum computing is the new kid on the block. Instead of relying on traditional bits of 1's and 0's, quantum computing relies on qubits that can exist in multiple states at once, making quantum computers many times more powerful than traditional ones. The potential applications of quantum computing are endless, from revolutionizing medicine to speeding up artificial intelligence.

Challenges to Quantum Computing Advancement

While the potential benefits of quantum computing are vast, some hurdles need to be overcome before it becomes a mainstream technology. One of the primary challenges is developing a scalable quantum system. Because quantum computers are so complex, building one that can be scaled to perform complex computations is still in its early stages. Additionally, the quantum computing field is highly competitive, with tech giants and startups alike racing to create the most powerful quantum computer.

Predictions for Quantum Computing

Despite the challenges, there is no doubt that quantum computing is on the rise. Experts predict that within the next few decades, quantum computers will become mainstream. Researchers are already making strides in developing quantum algorithms, and governments and investors are pouring money into the field. As quantum computers become more powerful and easier to use, the applications of the technology will only continue to grow.

Benefits and Challenges of Embracing the Quantumania World

Embracing the world of quantum computing offers many benefits, such as creating new opportunities for innovation and changing the way we approach problems. However, there are also challenges to consider, including security and ethical concerns.

Opportunities for Innovation

Quantum computing has the potential to revolutionize industries from healthcare to finance. Some of the applications currently being explored include drug discovery, financial modeling, and encryption. The development of new quantum algorithms will unlock new possibilities and change the way we approach complex problems.

Security and Ethical Considerations

As with any new technology, there are concerns about the security implications of quantum computing. For example, quantum computing could be used to break encryption much faster than traditional computers, which would be a significant cybersecurity threat. Additionally, there are ethical considerations to be aware of, such as the potential impact on privacy and the potential for job displacement.

Getting Started with Quantum Computing

If you're interested in learning more about quantum computing, there are many resources available to get you started.

Resources for Learning Quantum Computing

There are many online courses and tutorials available that cover the basics of quantum computing. Some popular resources include the IBM Quantum Experience, Microsoft Quantum, and the Quantum Computing Playground.

How to Get Involved in the Quantum Community

The quantum computing community is growing rapidly, and there are many opportunities to get involved. Attending conferences and joining online forums are great ways to connect with others in the field. Additionally, there are many open-source projects available for those who want to contribute to the development of the technology.

Conclusion: The Importance of Embracing the Quantumania World

The rise of quantum computing is bringing new possibilities and challenges. As we continue to explore the potential applications of quantum computing, it's essential to consider the security and ethical implications. With the right approach, we can unlock the full potential of this exciting new technology, creating a better world for everyone.

Final Thoughts on the Future of Quantum Computing

While the future of quantum computing is uncertain, one thing is clear: it has the potential to change the world as we know it. As researchers and developers continue to push the limits of what is possible, we can look forward to a future where complex problems can be solved faster and more efficiently than ever before. In conclusion, quantum computing is a rapidly evolving field with enormous potential to revolutionize various industries. While there are still many challenges to overcome, the benefits of embracing the Quantumania world cannot be understated. By learning about quantum computing, getting involved in the quantum community, and staying up-to-date with the latest advancements, individuals and organizations can position themselves at the forefront of this exciting technological revolution.

FAQ

What is quantum computing, and how does it differ from classical computing?

Quantum computing is a type of computing that uses the principles of quantum mechanics to process information. While classical computers use bits to store and process data (where each bit is either a 0 or 1), quantum computers use quantum bits (qubits) that can exist in various states simultaneously. This allows quantum computers to perform certain calculations much faster than classical computers.

What are some potential applications of quantum computing?

Quantum computing has the potential to revolutionize various industries, including finance, healthcare, energy, and transportation. Some potential applications include developing more secure cryptography, optimizing logistical systems, and accelerating the development of new drugs.

What challenges does quantum computing face?

One of the biggest challenges facing quantum computing is the issue of error correction. Quantum systems are extremely sensitive to noise and interference, which can cause errors in computations. Additionally, quantum systems are expensive and difficult to build, which has limited their widespread adoption.

How can I get involved in the quantum community?

There are many resources available for those interested in learning more about quantum computing and getting involved in the quantum community. Online forums and discussion boards, as well as conferences and workshops, are excellent ways to connect with other quantum enthusiasts and stay up-to-date with the latest advancements in the field.

As we move into an era of unprecedented technological advancements, the world of computing is on the cusp of a major transformation. Quantum computing, a revolutionary technology that has long been the stuff of science fiction, is becoming a reality. With its potential to solve complex problems at a speed that is unmatched by classical computers, quantum computing is poised to usher in a new era of innovation and progress. In this article, we will explore the basics of quantum computing, its advantages, challenges, and potential applications, and the ethical implications that arise as we move toward a quantum future.

1. Quantum Computing: The Future is Now

The Rise of Quantum Computing

In recent years, quantum computing has emerged as a new frontier in the field of computer science. While classical computers use bits to represent information, quantum computers use quantum bits, or qubits, to process information. The world of quantum computing is still in its early stages, but the potential for its applications is vast. From drug discovery to cryptography, quantum computing has the potential to revolutionize multiple industries.

1. What is Quantum Computing?

Understanding Quantum Mechanics and Quantum Bits (qubits)

How Quantum Computers Work

Quantum mechanics is a branch of physics that deals with the behavior of matter and energy at a quantum level. At this level, particles exhibit unusual properties like superposition, entanglement, and interference. Quantum computers are built on the principles of quantum mechanics and use qubits instead of classical bits. A qubit can be in two states simultaneously, allowing for multiple calculations to be performed simultaneously.

Quantum computers are built using superconductors, ions, or other systems that can maintain quantum states. These systems are kept at extremely low temperatures and isolated from the environment to prevent decoherence, which is the loss of quantum coherence. Quantum algorithms are designed to take advantage of these unique properties of qubits and perform calculations that are beyond the capabilities of classical computers.

1. The Advantages of Quantum Computing

Speed and Efficiency
Parallel Processing Capabilities
Quantum Cryptography and Security

Quantum computing has several advantages over classical computing. The most significant advantage is speed and efficiency. Quantum computers can perform calculations much faster than classical computers, allowing for complex simulations and modeling. Quantum computers can also perform parallel processing much more efficiently than classical computers, which can lead to faster data processing times.

Another significant advantage of quantum computing is in cryptography and security. Quantum cryptography is based on the principles of quantum mechanics and offers unparalleled security. The security of classical cryptography is based on the assumption that certain mathematical problems are difficult to solve. Quantum computers can solve these problems easily, rendering classical cryptography useless. Quantum cryptography, on the other hand, is based on the principles of quantum mechanics, making it impossible to break using classical computing methods.

1. The Quantum Computing Industry: Key Players and Developments

Major Players in the Quantum Computing Industry Recent Developments and Breakthroughs in Quantum Computing

The quantum computing industry is still in its nascent stages, but there are several key players in the market. IBM, Google, Microsoft, and Intel are some of the major players investing heavily in quantum computing. These companies are building quantum computers, developing quantum algorithms, and providing cloud-based quantum services.

In recent years, there have been several breakthroughs and developments in the field of quantum computing. In 2017, Google announced that its quantum computer could solve a complex problem in 200 seconds, which would take a classical computer 10,000 years. In 2019, IBM unveiled its first commercial quantum computer, the IBM Q System One. In 2020, Honeywell announced that it had built the world's most powerful quantum computer.

The future of quantum computing is exciting, and the potential for its applications is vast. From drug discovery to cryptography, quantum computing has the potential to revolutionize multiple industries. As more companies invest in this technology, we can expect significant developments and breakthroughs in the years to come.

1. Quantum Computing and Cybersecurity

The Threats of Quantum Computing to Current Cryptography

Most of our current cybersecurity systems rely on the fact that it is extremely difficult to factorize large numbers into their prime components. However, quantum computing's ability to perform complex calculations at an exponential rate threatens the security of these systems. Quantum computers can use Shor's algorithm to solve the encryption algorithms that are currently used to secure our online banking, email, and other sensitive data. This means that all the sensitive information we have online can be accessed by hackers with access to quantum computers.

Post-Quantum Cryptography and Security Solutions

To mitigate these risks, researchers have been working on developing post-quantum cryptography that can withstand the power of quantum computing. These new security protocols include lattice-based cryptography, code-based cryptography, hash-based cryptography, and multivariate cryptography. These new protocols offer promising alternatives that can operate with the same level of security as the current systems but with the added advantage of being resistant to quantum computing.

1. Ethical Implications of Quantum Computing

Impact on Employment and Job Displacement

The development of quantum computing is likely to lead to significant job displacement, particularly in industries reliant on computation and data analysis such as finance, healthcare, and logistics. As quantum computing becomes more prevalent, it may lead to the automation of many jobs that are currently carried out by humans. This raises serious ethical questions about the responsibility of governments and businesses to provide retraining and income support for those who become unemployed as a result of quantum computing.

Quantum Computing and AI: Opportunities and Risks

Quantum computing and AI will be able to work together to create revolutionary advances in many industries, including medicine and transportation. However, this partnership also poses risks, such as the potential for AI to gain access to sensitive data or for quantum computers to be used as tools for high-precision weapons. Ethical considerations must be taken into account during research and development to ensure that the potential harm is minimized.

1. Challenges Facing Quantum Computing

Technical Challenges and Limitations

Quantum computing is still in its early stages, and many technical challenges and limitations need to be overcome before it can be fully realized. One of the biggest challenges is the problem of noise, which can reduce the accuracy of quantum calculations. Additionally, the development of the infrastructure required to support quantum computing, such as cryogenic cooling systems, is still in progress.

Commercialization and Adoption

The commercialization and adoption of quantum computing within the business world, and society more broadly, is also a major challenge. As quantum computing is still in its early stages, the practical applications of the technology are not yet fully understood. Additionally, the cost of developing and operating quantum computers is still very high, making it difficult for smaller businesses to invest.

1. The Promise of a Quantum Future

Potential Applications of Quantum Computing

Quantum computing provides enormous potential for scientific research, particularly in the fields of physics, chemistry, and biology, where it can be used to simulate complex biological and chemical systems. It can also revolutionize energy systems, improve transport infrastructure, and improve weather forecasting. The potential applications of this technology are endless.

The Future of Quantum Research and Development

Quantum computing is still largely in its infancy, but the progress that has been made is impressive. The development of more effective and robust quantum computers is likely to continue well into the future, with new applications emerging as the technology develops. Furthermore, the interdisciplinary nature of quantum computing research offers exciting opportunities for new collaborations and cross-disciplinary approaches, leading to further advances in the field. As we conclude this exploration of the fascinating world of quantum computing, it is clear that this technology holds immense potential for the future. While there are certainly challenges that need to be overcome, we will likely see rapid advancements in this field in the coming years. With the promise of solving complex problems that are beyond the capabilities of classical computers, quantum computing presents a new frontier in the world of computing and technology. The future is indeed bright, and we can look forward to a world where quantum computing is at the forefront of innovation and progress.

Frequently Asked Questions

What is the difference between quantum computing and classical computing?

Quantum computing is based on the principles of quantum mechanics and uses qubits to perform calculations. Classical computing, on the other hand, is based on classical physics and uses bits. While classical computers can only perform one calculation at a time, quantum computers can perform multiple calculations simultaneously, making them much faster and more efficient.

What are the potential applications of quantum computing?

Quantum computing has the potential to revolutionize many fields, including cryptography, chemistry, medicine, finance, and logistics. For example, quantum computing could be used to develop new drugs, optimize supply chains, and solve complex optimization problems.

What are the challenges facing quantum computing?

One of the biggest challenges facing quantum computing is the issue of error correction. Quantum computers are highly sensitive to errors, and even small errors can result in incorrect results. Other challenges include scalability, cost, and the need for specialized infrastructure.

Chapter 10: A Quantum Future Beckons

Quantum computing raises several ethical questions, particularly around the impact on employment and the potential displacement of workers. There are also concerns about the use of quantum computing for military purposes and the potential for quantum computers to break current encryption methods and compromise cybersecurity. As we move into an era of unprecedented technological advancements, the world of computing is on the cusp of a major transformation. Quantum computing, a revolutionary technology that has long been the stuff of science fiction, is becoming a reality. With its potential to solve complex problems at a speed that is unmatched by classical computers, quantum computing is poised to usher in a new era of innovation and progress. In this article, we will explore the basics of quantum computing, its advantages, challenges, and potential applications, and the ethical implications that arise as we move toward a quantum future.

1. Quantum Computing: The Future is Now

The Rise of Quantum Computing

In recent years, quantum computing has emerged as a new frontier in the field of computer science. While classical computers use bits to represent information, quantum computers use quantum bits, or qubits, to process information. The world of quantum computing is still in its early stages, but the potential for its applications is vast. From drug discovery to cryptography, quantum computing has the potential to revolutionize multiple industries.

1. What is Quantum Computing?

Understanding Quantum Mechanics and Quantum Bits (qubits)

How Quantum Computers Work

Quantum mechanics is a branch of physics that deals with the behavior of matter and energy at a quantum level. At this level, particles exhibit unusual properties like superposition, entanglement, and interference. Quantum computers are built on the principles of quantum mechanics and use qubits instead of classical bits. A qubit can be in two states simultaneously, allowing for multiple calculations to be performed simultaneously.

Quantum computers are built using superconductors, ions, or other systems that can maintain quantum states. These systems are kept at extremely low temperatures and isolated from the environment to prevent decoherence, which is the loss of quantum coherence. Quantum algorithms are designed to take advantage of these unique properties of qubits and perform calculations that are beyond the capabilities of classical computers.

1. The Advantages of Quantum Computing

Speed and Efficiency
Parallel Processing Capabilities
Quantum Cryptography and Security

Quantum computing has several advantages over classical computing. The most significant advantage is speed and efficiency. Quantum computers can perform calculations much faster than classical computers, allowing for complex simulations and modeling. Quantum computers can also perform parallel processing much more efficiently than classical computers, which can lead to faster data processing times.

Another significant advantage of quantum computing is in cryptography and security. Quantum cryptography is based on the principles of quantum mechanics and offers unparalleled security. The security of classical cryptography is based on the assumption that certain mathematical problems are difficult to solve. Quantum computers can solve these problems easily, rendering classical cryptography useless. Quantum cryptography, on the other hand, is based on the principles of quantum mechanics, making it impossible to break using classical computing methods.

1. The Quantum Computing Industry: Key Players and Developments

Major Players in the Quantum Computing Industry Recent Developments and Breakthroughs in Quantum Computing

The quantum computing industry is still in its nascent stages, but there are several key players in the market. IBM, Google, Microsoft, and Intel are some of the major players investing heavily in quantum computing. These companies are building quantum computers, developing quantum algorithms, and providing cloud-based quantum services.

In recent years, there have been several breakthroughs and developments in the field of quantum computing. In 2017, Google announced that its quantum computer could solve a complex problem in 200 seconds, which would take a classical computer 10,000 years. In 2019, IBM unveiled its first commercial quantum computer, the IBM Q System One. In 2020, Honeywell announced that it had built the world's most powerful quantum computer.

The future of quantum computing is exciting, and the potential for its applications is vast. From drug discovery to cryptography, quantum computing has the potential to revolutionize multiple industries. As more companies invest in this technology, we can expect significant developments and breakthroughs in the years to come.

1. Quantum Computing and Cybersecurity

The Threats of Quantum Computing to Current Cryptography

Most of our current cybersecurity systems rely on the fact that it is extremely difficult to factorize large numbers into their prime components. However, quantum computing's ability to perform complex calculations at an exponential rate threatens the security of these systems. Quantum computers can use Shor's algorithm to solve the encryption algorithms that are currently used to secure our online banking, email, and other sensitive data. This means that all the sensitive information we have online can be accessed by hackers with access to quantum computers.

Post-Quantum Cryptography and Security Solutions

To mitigate these risks, researchers have been working on developing post-quantum cryptography that can withstand the power of quantum computing. These new security protocols include lattice-based cryptography, code-based cryptography, hash-based cryptography, and multivariate cryptography. These new protocols offer promising alternatives that can operate with the same level of security as the current systems but with the added advantage of being resistant to quantum computing.

1. Ethical Implications of Quantum Computing

Impact on Employment and Job Displacement

The development of quantum computing is likely to lead to significant job displacement, particularly in industries reliant on computation and data analysis such as finance, healthcare, and logistics. As quantum computing becomes more prevalent, it may lead to the automation of many jobs that are currently carried out by humans. This raises serious ethical questions about the responsibility of governments and businesses to provide retraining and income support for those who become unemployed as a result of quantum computing.

Quantum Computing and AI: Opportunities and Risks

Quantum computing and AI will be able to work together to create revolutionary advances in many industries, including medicine and transportation. However, this partnership also poses risks, such as the potential for AI to gain access to sensitive data or for quantum computers to be used as tools for high-precision weapons. Ethical considerations must be taken into account during research and development to ensure that the potential harm is minimized.

1. Challenges Facing Quantum Computing

Technical Challenges and Limitations

Quantum computing is still in its early stages, and many technical challenges and limitations need to be overcome before it can be fully realized. One of the biggest challenges is the problem of noise, which can reduce the accuracy of quantum calculations. Additionally, the development of the infrastructure required to support quantum computing, such as cryogenic cooling systems, is still in progress.

Commercialization and Adoption

The commercialization and adoption of quantum computing within the business world, and society more broadly, is also a major challenge. As quantum computing is still in its early stages, the practical applications of the technology are not yet fully understood. Additionally, the cost of developing and operating quantum computers is still very high, making it difficult for smaller businesses to invest.

1. The Promise of a Quantum Future

Potential Applications of Quantum Computing

Quantum computing provides enormous potential for scientific research, particularly in the fields of physics, chemistry, and biology, where it can be used to simulate complex biological and chemical systems. It can also revolutionize energy systems, improve transport infrastructure, and improve weather forecasting. The potential applications of this technology are endless.

The Future of Quantum Research and Development

Quantum computing is still largely in its infancy, but the progress that has been made is impressive. The development of more effective and robust quantum computers is likely to continue well into the future, with new applications emerging as the technology develops. Furthermore, the interdisciplinary nature of quantum computing research offers exciting opportunities for new collaborations and cross-disciplinary approaches, leading to further advances in the field. As we conclude this exploration of the fascinating world of quantum computing, it is clear that this technology holds immense potential for the future. While there are certainly challenges that need to be overcome, we will likely see rapid advancements in this field in the coming years. With the promise of solving complex problems that are beyond the capabilities of classical computers, quantum computing presents a new frontier in the world of computing and technology. The future is indeed bright, and we can look forward to a world where quantum computing is at the forefront of innovation and progress.

Frequently Asked Questions

What is the difference between quantum computing and classical computing?

Quantum computing is based on the principles of quantum mechanics and uses qubits to perform calculations. Classical computing, on the other hand, is based on classical physics and uses bits. While classical computers can only perform one calculation at a time, quantum computers can perform multiple calculations simultaneously, making them much faster and more efficient.

What are the potential applications of quantum computing?

Quantum computing has the potential to revolutionize many fields, including cryptography, chemistry, medicine, finance, and logistics. For example, quantum computing could be used to develop new drugs, optimize supply chains, and solve complex optimization problems.

What are the challenges facing quantum computing?

One of the biggest challenges facing quantum computing is the issue of error correction. Quantum computers are highly sensitive to errors, and even small errors can result in incorrect results. Other challenges include scalability, cost, and the need for specialized infrastructure.

What are the ethical implications of quantum computing?

Quantum computing raises several ethical questions, particularly around the impact on employment and the potential displacement of workers. There are also concerns about the use of quantum computing for military purposes and the potential for quantum computers to break current encryption methods and compromise cybersecurity.